YOUR KNOWLEDGE HAS VALUE

AF302681

- We will publish your bachelor's and
  master's thesis, essays and papers

- Your own eBook and book -
  sold worldwide in all relevant shops

- Earn money with each sale

Upload your text at www.GRIN.com
and publish for free

Bibliographic information published by the German National Library:

The German National Library lists this publication in the National Bibliography; detailed bibliographic data are available on the Internet at http://dnb.dnb.de .

This book is copyright material and must not be copied, reproduced, transferred, distributed, leased, licensed or publicly performed or used in any way except as specifically permitted in writing by the publishers, as allowed under the terms and conditions under which it was purchased or as strictly permitted by applicable copyright law. Any unauthorized distribution or use of this text may be a direct infringement of the author s and publisher s rights and those responsible may be liable in law accordingly.

Imprint:

Copyright © 2016 GRIN Verlag
Print and binding: Books on Demand GmbH, Norderstedt Germany
ISBN: 9783668545069

This book at GRIN:

https://www.grin.com/document/377051

Anthony Li

The Impact of a Seahorse's Size on the Mating Process. Is There a Significant Difference Between the Heights of Brooding Male Hippocampus Reidi as Compared to Non-Brooding Males?

GRIN Verlag

GRIN - Your knowledge has value

Since its foundation in 1998, GRIN has specialized in publishing academic texts by students, college teachers and other academics as e-book and printed book. The website www.grin.com is an ideal platform for presenting term papers, final papers, scientific essays, dissertations and specialist books.

Visit us on the internet:

http://www.grin.com/

http://www.facebook.com/grincom

http://www.twitter.com/grin_com

International Bacculaureate

Higher Level Biology – Internal Assessment

May 2016

Candidate Name: Anthony Li

School: Ashbury College

Subject Area: Biology

Title: Seahorses: Does Size Really Matter?

Research Question: Does Size Give Male *Hippocampus reidi* an Advantage in Finding Mates?

An Examination for the presence of a Significant Difference Between the Heights of Brooding Male Hippocampus reidi as compared to Non-Brooding Males

Word Count: 1881

Table of Contents

<u>PREFACE</u>

<u>Background Information</u>:

As a child, I had found seahorses the most particular out of all marine creatures. I was taught in elementary school that unlike with other species, male seahorses were the carriers of their offspring. I did not understand how a male organism could become pregnant, let alone give birth; that would be a question that would be stuck in my mind for a long time. A few years later, I would learn that female seahorses will deposit eggs into the male's brood pouch. (Fig. 1) (Danielson, 2002)

Fig. 1

Hippocampus erectus female transferring eggs to male. Retrieved from (National Aquarium, 2012) https://aqua.org/blog/2012/november/thoughtful-thursdays-chesapeake-bay-lined-seahorses

Although I now knew the process in which the male seahorse got pregnant, I began wondering what sort of advantage seahorses could have by having the males give birth. After researching this sexual role reversal, I discovered that scientists themselves had the same question with no concrete results. There were two potential reasons: the first was that since creating life takes quite a bit of energy, having the male seahorse carry the babies splits the energy load between the two parents. A second reason was that this allows seahorses to reproduce more quickly, since it gives the female seahorse more time to make more eggs; this is logical as seahorses have a low survival rate in the wild. (Texas A&M University, 2008) (Garrick-Maidment, 2004, p. 13). However, since seahorses can deliver up to 1500 eggs, but have an infant survival rate of merely 0.5%, I began wondering what made it so difficult to live to adulthood. (Kateman, 2011) (Danielson, 2002) One factor was that seahorses themselves are extremely fragile, sensitive to even the slightest changes in their environment. I also discovered that most seahorse species are endangered. Over the summer of 2015, I travelled to the west coast and visited the Vancouver Aquarium for the first time, where I saw many different marine animals including the seahorse species *Hippocampus erectus*. I was fascinated by how sluggish they seemed swimming in the tank; I had always thought that they could move faster. A month later, I travelled to the east coast, this time visiting the Huntsman Marine Science Centre. This time I saw a completely different set of marine species, but I was still drawn to the seahorses. The seahorses at Huntsman were *Hippocampus hippocampus*, a smaller and less common species of seahorse. Though it was a different species, I saw similarities to those that I saw in Vancouver; the same sluggish movement. The seahorses seemed to prefer to simply cling to vegetation and rest. I spent hours in front of the seahorse tank, examining their feeding and behavioural traits, and bombarding the guides there with questions. I was told that the seahorses had the ability to change colour, and that every morning they would be a pale white since all the lights in the aquarium were closed at 5pm every night. I was doubtful, but after standing for a few hours with a red coat, I noticed that the seahorses had a noticeable red tinge (Fig. 2).

When I found out that they were captive bred, I started wondering what differences there were between captive bred seahorses and seahorses from the wild. I pondered the

Fig. 2

Photocredit: Tony Li, September 2nd, 2015: A *Hippocampus hippocampus* rests by hanging
onto a blade of grass by its tail

differences in size, aggressiveness and feeding habits among other variables. With all the

different possibilities, I was excited to begin research. However, when I began searching for

data online, I found very little data on seahorses. Since most seahorse species are endangered,

there is very little data on them in the wild due to their fragility and conservation status. I

quickly realized that I would not be able to find data comparing the seahorse infant mortality

rate in the wild as opposed to in captivity, due to the environmental and ethical

considerations that have been agreed upon by the global scientific community. (Morgan,

2013, p. 6) After extensive research about seahorses through databases, I found detailed data

on *Hippocampus reidi* concerning the heights of brooding and non-brooding males, and

mature and immature females. The data reminded me of my younger interest on male

seahorses giving birth, and thus I decided to see if there was a significant height difference

between brooding male seahorses as opposed to non-brooding males. The measuring of

seahorse height is a quick process, and has not been believed to do any harm for years

(Lourie, 2003, p. 5).

<u>INTRODUCTION</u>

<u>**Research Question:**</u>

Does Size Give Male *Hippocampus reidi* an Advantage in Finding Mates?

The term "brooding male" refers to male *Hippocampus reidi* which were observed to having a bulging brooding pouch containing seahorse eggs. Wild seahorses breed only once a year during their respective breeding seasons. (Texas A&M University, 2008) This allows one to assume that brooding males had found mates and were carrying their offspring, while non-brooding males had not found mates and therefore did not have any offspring to carry. The brooding pouch extends horizontally from seahorses, meaning that its presence will not affect height. In the wild, size can be quite important. Larger male seahorses can intimidate smaller seahorses, especially when competing for a mate. A bigger sized seahorse likely can defend its territory and offspring better than a smaller sized seahorse, thus becoming more appealing for female seahorses. In addition, a larger seahorse could signify an older, more experienced and capable seahorse.

<u>BODY</u>

<u>Hypothesis:</u>

I believe that there will be a significant difference between the heights of brooding

male and non-brooding male *Hippocampus reidi*, since I believe larger male seahorses would

be more successful breeding mates.

<u>Variables:</u>

The independent variable is comparing brooding males to non-brooding males, while

the dependent variable is the mean height of each seahorse group. The unpaired T-Test will

be conducted to examine how significant a difference there is between the heights of

brooding male seahorses and non-brooding male seahorses. The data was found from a study

done in 2007 by the corporation Sociedade Brasileira de Ictiologia, in which 911 seahorses

were recorded: 322 adult males (35.35%), 283 adult females (31.06%) and 306 juveniles

(33.60%), resulting in an unbiased sex ratio. The data can be found in the following paper:

https://www.academia.edu/3855631/Population_characteristics_space_use_and_habitat_asso

ciations_of_the_seahorse_Hippocampus_reidi_Teleostei_Syngnathidae (Lucena et al., 2007,

p. 4). In terms of controlled variables, no experimental procedures were done and so the only

controls are those that the researchers put in place. The researchers gathered data using a 50m

x 2m transect line along the NE, SE and S portions of the Brazilian coast. The visual censuses

were done between September of 2002 and September of 2004, however, additional data was

also collected between October of 2004 and September of 2006. Data was collected during

the morning or afternoon only, through use of snorkels or SCUBA gear. In terms of the data

collected on male seahorses, the researchers determined sexual maturity based on how developed the pouches were. All seahorse specimens were either photographed, filmed, or

marked around the tail region with a cotton rib.

<u>Raw Data:</u>

Surveyed locality (Brazilian state)	N° Seahorses sighted (N° transects)	Height of non-brooding males	Height of brooding males
Camurupim (Piauí)	125 (47)	14.7 ± 0.6 (11.0-17.0)	14.8 ± 0.5 (12.5-17.5)
Ubatuba (Piauí)	9 (18)	13.0	13.0
Pacoti (Ceará)	169 (80)	8.5 ± 0.3 (4.2-13.5)	11.6 ± 2.8 (6.5-16.4)
Mal Cozinhado (Ceará)	62 (19)	3.5	11.0
Tubarão (Rio Grande de Norte)	134 (26)	15.0 ± 0.5 (10.5-17.2)	15.9 ± 0.5 (13.5-18.0)
Casqueira (Rio Grande de Norte)	198 (56)	15.6 ± 0.5 (11.0-19.0)	16.6 ± 0.4 (14.0-20.0)
Mamanquape (Paraíba)	41 (47)	13.25 ± 2.4 (11.0-15.0)	14.9 ± 1.1 (12.0-18.0)
Itapessoca (Pernambuco)	80 (57)	14.5 ± 0.43 (10.4-16.0)	14.7 ± 0.5 (13.1-15.8)
Ariquindá (Pernambuco)	42 (53)	16.6 ± 0.8 (12.0-18.0)	16.1 ± 2.36 (15.1-17.0)
Andorinhas (Rio de Janeiro)	19 (19)	12 ± 0.7 (7.5-12.3)	11.6; 13.7
Itaipu (Rio de Janeiro)	27 (6)	14.7 ± 3.4 (9.1-16.1)	16.0 ± 1.3 (12.6-19.0)

Adapted from: Lucena et al., 2007, p. 4

An unpaired t-test will be used since the data is continuous, and there are two means which have a difference. Since there are two treatment groups, but the measurements in each treatment do not affect one another, an unpaired t-test will be used. It should be noted that a t-test assumes a normal distribution curve for the data.

Calculations:

H_o: There will be no significant difference between the mean heights of brooding males as opposed to the mean height of non-brooding males.

H_A: There will be a significant difference between the mean heights of brooding males as opposed to the mean height of non-brooding males.

d.f. : $(n_1+n_2) - 2$

$$=11 + 11 - 2$$

$$=20$$

$\bar{x}$ height of non-brooding males = (14.7 + 13.0 +8.5 +3.5 + 15.0 + 15.6 + 13.25 + 14.5 + 16.6

+ 12 + 14.7)/11

$\cong$ 12.8 cm $\pm$ 0.9 cm

Uncertainty calculation = (0.6 + 0 + 0.3 + 0.5 + 0 + 0.5 + 2.4 + 0.43 + 0.8 + 0.7 + 3.4)/11

$\cong$ 0.9 cm

Standard deviation of non-brooding male height $\cong$ 3.78

$\bar{x}$ height of brooding males = 14.3 cm $\pm$ 0.9 cm

Uncertainty calculation $\cong$ 0.9 cm

Standard deviation of brooding male height $\cong 1.93$

$$t = \frac{\overline{x_1} - \overline{x_2}}{\sqrt{\dfrac{S_1^{\,2}}{N_1} + \dfrac{S_2^{\,2}}{N_2}}}$$

$$t \cong \frac{|12.8 - 14.3|}{\sqrt{\left(\dfrac{3.78^2}{11} + \dfrac{1.93^2}{11}\right)}}$$

$$t \cong \frac{|-1.5|}{1.2797}$$

$$t \cong 1.17$$

Uncertainty: ± 1.4

Critical t-value at 5% for d.f. of 20 is 2.086.

Since 2.086 > 1.17,

Critical t-value > calculated t-value

<u>CONCLUSION/EVALUATION</u>

Therefore, since the calculated t-value of 1.13 is lower than than the critical t-value of 2.086 at the 95% confidence level, we accept the null hypothesis; there is no significant difference between the mean heights of brooding male *Hippocampus reidi* as opposed to the mean height of non-brooding male *Hippocampus reidi*.

The data in the calculations does not agree with the initial hypothesis made, which said that there would be a significant difference between the mean heights since I believed that larger male seahorses would be more successful breeding mates. The fact that there is not a significant difference hints at other variables being involved in the *Hippocampus reidi* mating ritual. These could be skin colour vibrancy, swimming speed or ability to catch prey.

There were a few potential areas of improvement. First and foremost, there was an error on the final t-value of 1.4. This error put the t-value both over the critical t-value, and far under it. The gigantic error was primarily due to the variances in sample size. For instance, certain locations where seahorse data was collected had a small sample size, and were less representative of their location. Two of the locations for non-brooding males had one mere data point, and thus likely skewed the final result. Also, seahorse height could have been affected by a multitude of factors such as diet, age or even injury. Though these factors are difficult to control, perhaps male seahorses could be tagged and monitored over a period of time to see if their respective heights change at all when there are brooding. This would control their own individual factors such as diet, as they would most likely stay in the same area. If it would be noticed that the seahorses tagged continue to grow while non-brooding, it would be indicated that the seahorses have not reached their complete height. Unfortunately, with seahorses being an endangered species, additional data will be hard to come by. This investigation could continue over another span of a few years so that more data could be compiled and a more accurate conclusion could be reached.

REFERENCES

Danielson, S. (2002, June 14). Seahorse Fathers Take Reins in Childbirth. Retrieved October 27, 2015, from National Geographic website: http://news.nationalgeographic.com/news/2002/06/0614_seahorse_recov.html

Garrick-Maidment, N. (2004). *British Seahorse Survey Report 2004*. Retrieved from http://www.theseahorsetrust.org/userfiles/PDF/BSS%202004%20Report.pdf

Kateman, B. (2011, September 16). The Male Seahorse – Supermom? Retrieved October 31, 2015, from State of the Planet, Earth Institute, Columbia University website: http://blogs.ei.columbia.edu/2011/09/16/male-seahorse-a-supermom/

Lourie, S. (2003, May). Measuring Seahorses (Technical Report No. 4). Retrieved from http://static1.squarespace.com/static/55930a68e4b08369d02136a7/t/5639576ce4b05b184b56cb40/1446598508315/Measuring_Seahorses.pdf

Lucena, I., R., T. P., C., A. L., de Souza, L. E., A., J. H., & C., M. (2007). *Neotropical Ichthyology: Population characteristics, space use and habitat associations of the seahorse Hippocampus reidi.* http://dx.doi.org/10.1590/S1679-62252007000300020

Morgan, R. (2013, September). Maritime Infrastructure and the Protected Seahorse. *The Environmental Practitioner*, (31), 6. Retrieved from http://www.eianz.org/document/item/2860

National Aquarium. (2012, November 1). [Lined seahorses vary in color, pattern and ornamentation]. Retrieved from https://aqua.org/blog/2012/november/thoughtful-thursdays-chesapeake-bay-lined-seahorses

Texas A&M University. (2008, May 2). Male Seahorses Are Nature's Mr. Mom, Researchers Say. Retrieved November 1, 2015, from ScienceDaily website: http://www.sciencedaily.com/releases/2008/05/080501125451.htm

YOUR KNOWLEDGE HAS VALUE

- We will publish your bachelor's and master's thesis, essays and papers

- Your own eBook and book - sold worldwide in all relevant shops

- Earn money with each sale

Upload your text at www.GRIN.com and publish for free